Determination of the $\delta^{34}S$ of Total Sulfur in Solids; RSIL Lab Code 1800

By Kinga Révész, Haiping Qi, and Tyler B. Coplen

Chapter 4 of
Section C, Stable Isotope-Ratio Methods
Book 10, Methods of the Reston Stable Isotope Laboratory

Edited by Kinga Révész and Tyler B. Coplen

Techniques and Methods 10–C4

U.S. Department of the Interior
U.S. Geological Survey

U.S. Department of the Interior
KEN SALAZAR, Secretary

U.S. Geological Survey
Marcia K. McNutt, Director

U.S. Geological Survey, Reston, Virginia: 2012
Version 1.0, 2006
Version 1.1, 2007
Version 1.2, 2012

For sale by U.S. Geological Survey, Information Services
Box 25286, Denver Federal Center
Denver, CO 80225

For more information about the USGS and its products:
Telephone: 1–888–ASK–USGS
World Wide Web: http://www.usgs.gov/

Suggested citation:
Révész, Kinga, Qi, Haiping, and Coplen, T.B., 2012, Determination of the $\delta^{34}S$ of total sulfur in solids; RSIL lab code 1800, chap. 4 *of* Stable isotope-ratio methods, sec. C *of* Révész, Kinga, and Coplen, T.B. eds., Methods of the Reston Stable Isotope Laboratory (slightly revised from version 1.1 released in 2007): U.S. Geological Survey Techniques and Methods, book 10, 31 p., available only at http://pubs.usgs.gov/tm/2006/tm10c4/. (Supersedes versions 1.0 and 1.1 released in 2006 and 2007, respectively.)

Foreword

The Reston Stable Isotope Laboratory (RSIL) provides stable isotope analyses on a routine basis for a large user community within the U.S. Geological Survey (USGS) and elsewhere. The RSIL also serves the USGS National Research Program (NRP) through its project on Stable Isotope Fractionation in Hydrologic Processes. The NRP conducts basic and problem-oriented hydrologic research in support of the mission of the USGS. The stable isotope project conducts research on the use of isotope-ratio measurements in studies of water resources and environmental quality. One objective of this project is to develop new techniques for isotopic analysis of hydrogen, nitrogen, oxygen, carbon, and sulfur in environmental samples. New analytical techniques expand the range of tools available for studying the movement of those elements in hydrologic and biogeochemical systems. Another objective of the project is to test new applications of isotope measurements in specific field settings. Field studies of isotope behavior have contributed to understanding water-supply sustainability, groundwater/surface-water interactions, paleoclimate history, biologic cycling of nutrients, groundwater contamination, and natural remediation. This project also contributes to the improvement of measurement science and the development of isotope databases.

Book 10 of the Techniques and Methods Series of the USGS contains detailed descriptions of methods of the RSIL, including routine analytical methods called standard operating procedures (SOPs), along with safety guidelines, maintenance procedures, and other information about the operation of the RSIL. Section C of book 10 contains the SOPs for a variety of methods to measure stable isotope ratios, each of which constitutes a chapter. Each chapter is limited to a narrow field of subject matter to permit flexibility in revision as the need arises.

Picrrc Glynn
Chief, Branch of Regional Research, Eastern Region

Contents

Figures

Conversion Factors

Multiply	By	To Obtain
Length		
millimeter (mm)	0.03937	inch
centimeter (cm)	0.3937	inch
meter (m)	3.281	foot
Volume		
micro liter (μL)	0.06102×10^{-3}	cubic inch
milliliter (mL)	0.06102	cubic inch
cubic centimeter (cm^3)	0.06102	cubic inch
Mass		
nanogram (ng) = 10^{-3} μg	3.527×10^{-11}	ounce
microgram (μg) = 10^{-3} mg	3.527×10^{-8}	ounce
milligram (mg) = 10^{-3} g	3.527×10^{-5}	ounce
gram (g) = 10^{-3} (kg)	3.527×10^{-2}	ounce
kilogram = 10^3 g; 10^6 mg; 10^9 μg; 10^{12} ng	2.205	pound
Density		
gram per cubic centimeter (g/cm^3)	0.5780	ounce per cubic inch
Temperature		
Celsius (°C)	°F = 9/5 (°C) + 32	Fahrenheit (°F)
Pressure (force/area)		
kilopascal (kPa)	9.869×10^{-3}	atmosphere, standard (atm)
kilopascal (kPa)	1.450×10^{-1}	pound-force/square inch (psi)
kilopascal (kPa)	1.000×10^{-2}	bar
kilopascal (kPa)	2.961×10^{-1}	inches of mercury at 60 °F

Acronyms and Abbreviations

ANSI	American National Standards Institute
CF-IRMS	continuous flow isotope-ratio mass spectrometer
cm	centimeter
DIW	deionized water
EA	elemental analyzer
GC	gas chromatograph
IAEA	International Atomic Energy Agency
IRMS	isotope-ratio mass spectrometer
kPa	kilopascal
LIMS-LSI	Laboratory Information Management System for Light Stable Isotopes
mg	milligram
mg/g	milligram per gram
mL	milliliter
mL/min	milliliter per minute
min	minute
MSDS	Material Safety Data Sheets
NBS	National Bureau of Standards
NFPA	National Fire Protection Association
NWQL	National Water-Quality Laboratory (USGS)
per mil, ‰	one part in one thousand parts, with value 10^{-3}
pF	picofarad
QA	quality assurance
QC	quality control
QC/QA	quality control/quality assurance
RSIL	Reston Stable Isotope Laboratory
SOP	standard operating procedure
std	international measurement standard
USGS	U.S. Geological Survey
VCDT	Vienna Cañon Diablo Troilite

Symbols

Ω	ohm
$<$	less than
$\leq$	less than or equal

Determination of the δ^{34}S of Total Sulfur in Solids; RSIL Lab Code 1800

By Kinga Révész, Haiping Qi, and Tyler B. Coplen

Summary of Procedure

The purpose of the Reston Stable Isotope Laboratory (RSIL) lab code 1800 is to determine the $\delta(^{34}S/^{32}S)$, abbreviated as δ^{34}S, of total sulfur in a solid sample. A Carlo Erba NC 2500 elemental analyzer (EA) is used to convert total sulfur in a solid sample into SO_2 gas. The EA is connected to a continuous flow isotope-ratio mass spectrometer (CF-IRMS), which determines the relative difference in stable sulfur isotope-amount ratio ($^{34}S/^{32}S$) of the product SO_2 gas. The combustion is quantitative; no isotopic fractionation is involved. Samples are placed in tin capsules and loaded into a Costech Zero-Blank Autosampler on the EA. Under computer control, samples are dropped into a heated reaction tube that combines both the oxidation and the reduction reactions. The combustion takes place in a helium atmosphere that contains an excess of oxygen gas at the oxidation zone at the top of the reaction tube. Combustion products are transported by a helium carrier through the reduction zone at the bottom of the reaction tube to remove excess oxygen and through a separate drying tube to remove any water. The gas-phase products, mainly CO_2, N_2, and SO_2, are separated by a gas chromatograph (GC). The gas is then introduced into the isotope-ratio mass spectrometer (IRMS) through a Thermo-Finnigan ConFlo II interface, which also is used to inject SO_2 reference gas and helium for sample dilution. The IRMS is a Thermo Delta V Plus CF-IRMS. It has a universal triple collector with two wide cups and a narrow cup in the middle. It is capable of measuring mass/charge (m/z) 64 and 66 simultaneously. The ion beams from SO_2 are as follows: m/z 64 = SO_2 = $^{32}S^{16}O^{16}O$; and m/z 66 = SO_2 = $^{34}S^{16}O^{16}O$ primarily.

Reporting Units and Operational Range

Variations in stable isotope ratios typically are small. Stable isotope ratios commonly are measured and expressed as the relative difference in the ratio of the number of the less abundant isotope (usually the heavy isotope) to the number of the more abundant isotope (usually the light isotope) of a sample with respect to the measurement standard, std (Coplen, 2011). This relative difference is designated δ^iE, which is a shortened form of $\delta(^iE/^jE)$ or $\delta(^iE)$, and is defined according to equation 1 (Coplen, 2011):

$$\delta^i E = \delta\left(^i E\right) = \delta\left(^i E/\,^j E\right) = \frac{N\left(^i E\right)_P / N\left(^j E\right)_P - N\left(^i E\right)_{std} / N\left(^j E\right)_{std}}{N\left(^i E\right)_{std} / N\left(^j E\right)_{std}} \tag{1}$$

where $N(^iE)_P$ and $N(^jE)_P$ are the numbers of the two isotopes ^iE and ^jE of element E in specimen P and equivalent parameters follow for the international measurement standard, "std." A positive δ^iE value indicates that the specimen is enriched in the heavy isotope, ^{i}E, relative to the std. A negative δ^iE value indicates that the specimen is depleted in the heavy isotope, ^{i}E, relative to the std. For stable sulfur isotope ratios, δ^{34}S is defined as follows:

$$\delta^{34}S \;=\; \delta\left(^{34}S\right) \;=\; \delta\left(^{34}S/^{32}S\right) \;=\; \frac{N\left(^{34}S\right)_P / N\left(^{32}S\right)_P - N\left(^{34}S\right)_{std} / N\left(^{32}S\right)_{std}}{N\left(^{34}S\right)_{std} / N\left(^{32}S\right)_{std}} \qquad\qquad \textbf{(2)}$$

The international measurement standard for measurement of the relative difference in sulfur isotope-number ratios ($\delta^{34}S$) is IAEA-S-1, which has a consensus $\delta^{34}S$ value of –0.3 ‰ relative to Vienna Cañon Diablo Troilite (VCDT) (Robinson, 1995; Krouse and Coplen, 1997; Coplen and Krouse, 1998). By interspersing isotopic reference materials with accepted $\delta^{34}S$ values among unknown samples, $\delta^{34}S$ values can be determined relative to VCDT. These isotope ratios are made with a CF-IRMS, which measures the ratios of the sample SO_2 gas and one or more injections of the "working reference" SO_2 gas.

The system was tested by analyzing pure inorganic samples ($BaSO_4$ and Ag_2S), an array of biological tissues from vertebrates and invertebrates, and geologic samples, such as suspended organic matter in sediment. The routine analysis requires a minimum of 40 micrograms (µg) of sample as sulfur in a maximum of 40 milligrams (mg) of solid sample; however, the system is capable of analyzing samples as small as 19 µg of sulfur in one sample aliquot with reduced analytical precision. The $\delta^{34}S$ values ranged from approximately –34 to +21 ‰, which covers the range of $\delta^{34}S$ values of most natural samples. It was found that the precision and accuracy of the results were acceptable (± 0.2 ‰), and no memory effects were observed.

Reference Materials and Documentation

Reference Materials Used, Storage Requirements, and Shelf Life

The primary reference material for $\delta^{34}S$ measurements is IAEA-S-1, which has a consensus $\delta^{34}S$ value of –0.3 ‰ relative to VCDT (Robinson, 1995; Krouse and Coplen, 1997; Coplen and Krouse, 1998). Other internationally distributed isotopic reference materials that are used have a wide range of $\delta^{34}S$ values. In the procedure reported herein, barium sulfates ($BaSO_4$) that are used for isotopic analysis of inorganic samples include NBS 127 with $\delta^{34}S$ of +21.1 ‰, IAEA-SO-5 with $\delta^{34}S$ of +0.5 ‰, and IAEA-SO-6 with $\delta^{34}S$ of –34.05 ‰. Silver sulfide (Ag_2S) reference materials are used for organic sulfur or sulfide samples, and they include IAEA-S-1 with $\delta^{34}S$ of –0.3 ‰, IAEA-S-2 with $\delta^{34}S$ of +21.1 ‰, and IAEA-S-3 with $\delta^{34}S$ of –32.55 ‰. No locally prepared isotopic reference materials are used in this method.

All of these reference materials are stored in the RSIL at ambient temperature in glass bottles capped with Teflon-coated or cone-shaped caps to keep moisture out. Their shelf life is indefinite.

Documentation

All calibration results are stored in the Laboratory Information Management System for Light Stable Isotopes (LIMS-LSI) (Coplen, 2000) under the following sample identifiers: S-1304 (IAEA-SO-5); S-97 (NBS 127); S-1302 (IAEA-SO-6); S-95 (IAEA-S-1); S-200 (IAEA-S-2); and S-808 (IAEA-S-3).

Labware, Instrumentation, and Reagents

Preparatory labware and apparatus include tin capsules and a microbalance capable of measuring samples with 0.001-mg precision. The analytical apparatus consists of four different units: (1) EA, (2) ConFlo interface, (3) IRMS, and (4) computer software.

The EA is a Carlo Erba Instruments NC 2500 system (CE NC 2500), and it has a Costech "Zero Blank" autosampler that holds 49 samples enclosed in tin capsules (fig. 1). Each tin capsule falls into the reaction tube, which is kept at 1020 °C and is under a constant helium flow (90 milliliters per minute (mL/min)). The sample immediately reacts with a measured amount of oxygen released from a 5-milliliter (mL) loop purged at 35 mL/min. The reaction of oxygen with the tin capsule is exothermic, resulting in localized temperatures of up to 1800 °C, thus ensuring complete and instantaneous sample combustion. This is called "dynamic flash combustion." Because the S-bearing solid has a decomposition temperature commonly in excess of 1500 °C, it is critical to synchronize the helium flow, the oxygen flow, and the timing of the sample drop to achieve quantitative combustion (Révész, 1998; also see in the present report the section titled "Troubleshooting and Bench Notes"). To boost the combustion, V_2O_5 is added to each sample (Yanagisawa and Sakai, 1983). The ratio of V_2O_5/S is not critical; usually the mass ratio is approximately 10. The combustion gases are first carried by helium through an oxidative catalyst layer (tungsten oxide, WO_3, on alumina) in the top of the reaction tube, where oxidation is completed, and then through a reduction agent (Cu) packed on the bottom of the reaction tube where any excess oxygen gas is consumed. At the outlet of the reaction tube, the gas mixture (N_2, CO_2, SO_2, and H_2O) enters a trap containing Anhydrone that absorbs water. The gas mixture, now N_2, CO_2, and SO_2, flows through a Teflon tubing GC column (0.8-m Porapak Q kept at 90 °C), which separates the gases and ensures that pure sulfur dioxide gas passes separately through the thermal conductivity detector, through the ConFlo II open split, and into the IRMS.

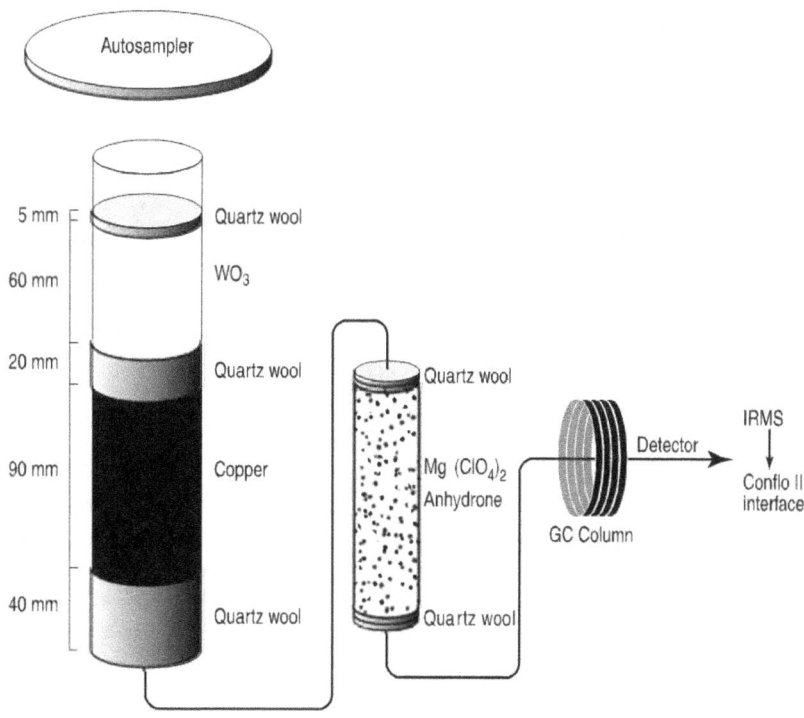

Figure 1.   Diagram of the Carlo Erba Instruments NC 2500 (CE NC 2500) elemental analyzer (used with permission of Thermo Electron Co.).

3

The Thermo-Finnigan ConFlo II provides the means for coupling the EA to the CF-IRMS with an open-split arrangement (fig. 2). The principle of an open split is that it reduces the 90 mL/min helium flow needed to operate the EA to a 0.3 mL/min helium flow needed to operate the IRMS. This is placing the end of the capillary that leads to the IRMS directly in the flow of helium coming from the EA. The narrow diameter and length of this capillary limit the flow of gas to the IRMS. Furthermore, extra capillaries carrying reference gas or helium gas can be added to the open split, thereby making it possible to inject reference gas to the IRMS or to dilute the sample with extra helium if the sample is too large (or to direct gases away by dilution if they are too large to ensure total separation).

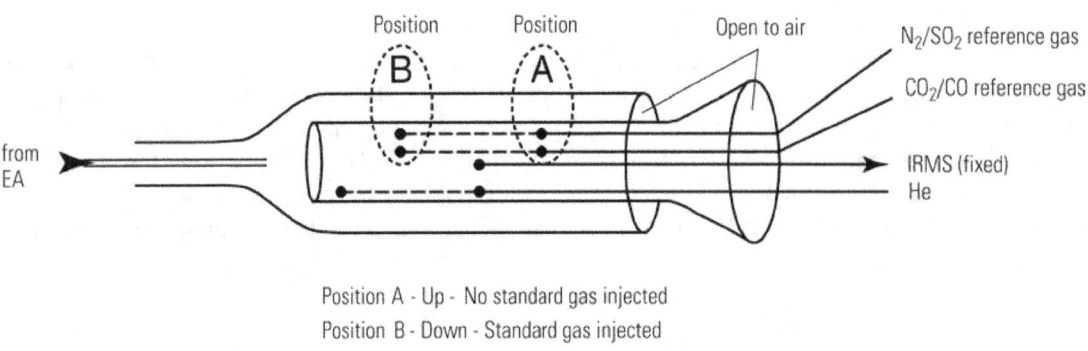

Position A - Up - No standard gas injected
Position B - Down - Standard gas injected

Figure 2. Diagram of a Finnigan ConFlo II Open Split (used with permission of Thermo Electron Co.).

The IRMS is a Thermo Delta V Plus CF-IRMS (fig. 3). The fundamental principle of the CF-IRMS technique is that a carrier gas transports the analyte through an initial stage of online chemistry for conversion to a form acceptable by the IRMS (Brenna and others, 1997). This is an automated system generating online, high-precision δ values of bulk solid and nonvolatile liquid samples. In the IRMS, gas molecules are ionized in a source by electrons emitted from a hot filament. The ions are accelerated into an analyzer, separated in a magnetic field, and collected in Faraday cup collectors. The ion beam intensities are measured with electrometers. This IRMS has a universal triple collector with two wide cups and a narrow cup in the middle. The resistor-capacitor combination on the electrometer of the low-mass cup is 3×10^8 ohms (Ω) and 680 picofarads (pF). At the high-mass cup, it is 3×10^{10} Ω and 5 pF. The instrument is capable of measuring m/z 64 and 66 simultaneously.

The Thermo Scientific ISODAT 2.0 software is designed (1) to advance the autosampler carousel, (2) to control the ConFlo II interface to inject reference gases at the desired time and dilute the sample if desired, (3) to operate the IRMS, (4) to acquire data from the IRMS, and (5) to calculate delta values.

Sample Collection, Preparation, Analysis, Retention Times, and Disposal

Sample Container, Preservation, and Handling Requirements

Each sample is collected in a high-density polyethylene scintillation vial with Polyseal cap (Wheaton 986706), provided by RSIL free of charge. Sample size should be a maximum of 1 gram (g). The sample should be homogenized, oven dried (90 °C), and powdered by the submitter to 100 to 200 micrometers (μm). Containers are labeled with isotopes whose concentrations are to be determined and labeled with the respective laboratory code or schedule number. A minimum of 0.15 mg of sulfur per sample is required. The minimum sample concentration is 1 milligram per gram (mg/g) sample for sulfur. Lower concentrations can be analyzed; contact the laboratory for further information. No treatment, preservation, or special shipping is required.

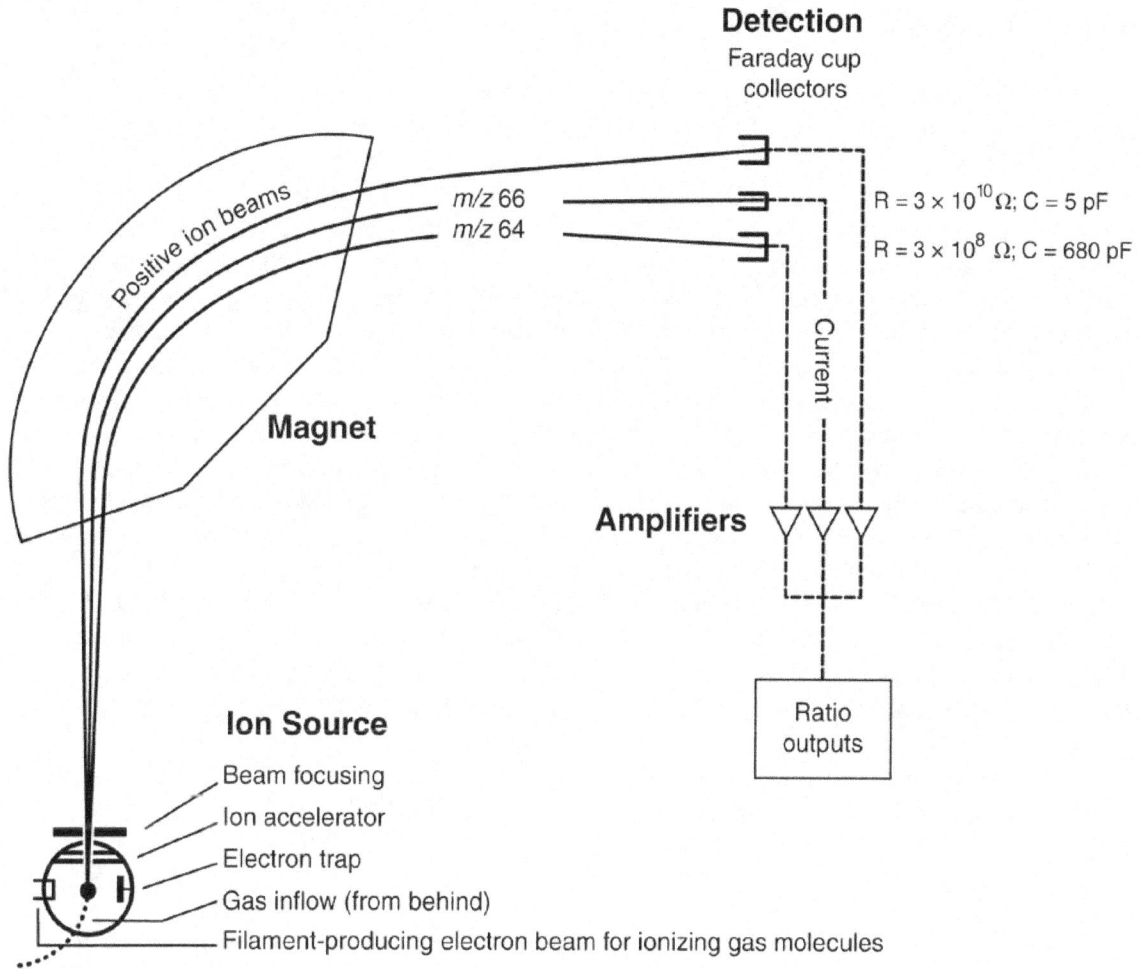

Detection
Faraday cup collectors

$R = 3 \times 10^{10}\,\Omega; C = 5\ pF$

$R = 3 \times 10^{8}\,\Omega; C = 680\ pF$

m/z 66
m/z 64

Positive ion beams

Magnet

Current

Amplifiers

Ratio outputs

Ion Source

Beam focusing
Ion accelerator
Electron trap
Gas inflow (from behind)
Filament-producing electron beam for ionizing gas molecules

Figure 3.   Schematic of a continuous flow isotope-ratio mass spectrometer (CF-IRMS) (modified from Clark and Fritz, 1997).

Sample Preparation and Time Requirements

Sample preparation involves logging samples in to the LIMS-LSI (see appendix A), creating a sample work-order sheet, and weighing samples into capsules.

The logging-in procedure involves logging samples in batches to LIMS-LSI that could contain from 1 to 100 samples per batch (also called a project), printing labels for each sample, and printing a summary project report. Each sample label contains the Field ID (the identifier by which the sample submitter knows each sample) and the unique RSIL Lab ID assigned by the LIMS-LSI. The summary project report is inserted into a three-ring binder titled "Samples in Progress."

An Excel sample workbook is generated. The step-by-step procedures are shown in appendix B. Sample weighing includes balance conditioning, weighing the appropriate amount of sample into the capsule, recording weight on the template, adding approximately 600 µg of V_2O_5 to the capsule, and closing the capsule by folding it. Step-by-step procedures are given in appendix C. The time requirement for sample preparation is approximately 1 to 5 days.

Performing the Analysis and Time Requirements

All samples are prepared and analyzed in batches of a maximum of 80 per day, and all samples are analyzed in duplicate. Six aliquots of three internationally distributed isotopic reference materials with different $\delta^{34}S$ values are analyzed at the beginning of a batch. Four of them are depleted in ^{34}S, two are enriched in ^{34}S, and one has an intermediate value. A set of these three reference materials is also interspersed among the samples. At least one reference material set is analyzed for every 10 unknowns. The amount of sulfur in the reference materials should be in the same range as that of the samples. Therefore, every sample must be preanalyzed to obtain its sulfur concentration unless it is provided by the customer. If the sulfur concentration is unknown and cannot be determined before isotopic analysis, a sequence of reference materials containing different amounts of sulfur must be analyzed to determine a mass-dependent correction factor. The carbon or nitrogen concentration also may be necessary to apply the correct dilution with helium to the CO_2 or N_2 to avoid peak interference of the SO_2 peak.

The procedure involves loading the autosampler carousel with capsules that contain an appropriate amount of sample in sequence according to the "Template" (appendix D), loading a new sample heading (including sample weights) into the "Sequence Table" in ISODAT 2.0 of the IRMS software, choosing the appropriate "Method," and starting sequence acquisition (see appendix E). After the analyses are completed, the data are exported (appendix F) to a Zip disk or "memory stick" from the ISODAT 2.0 database and transferred (appendix G) to the LIMS-LSI and the Data Back-Up computer. Final daily correction factors are determined with the LIMS-LSI (appendix H), based on the daily analyses of reference materials using equations 3, 4, and 5, which are presented with example numerical values:

$$-34.05 \text{ \textperthousand} = m \times \delta^{34}S_{\text{IAEA-SO-6/workingrefgas}} + b \tag{3}$$

$$+0.50 \text{ \textperthousand} = m \times \delta^{34}S_{\text{IAEA-SO-5/workingrefgas}} + b \tag{4}$$

$$+21.10 \text{ \textperthousand} = m \times \delta^{34}S_{\text{NBS127/workingrefgas}} + b \tag{5}$$

The values −34.05 ‰, +0.50 ‰, and +21.10 ‰ are the assigned values of the international reference materials IAEA-SO-6, IAEA-SO-5, and NBS 127, respectively. The $\delta^{34}S_{\text{ref/workingrefgas}}$ values in equations 3, 4, and 5 are the mean daily delta values of the respective international reference materials relative to the working reference gas; b is the additive correction factor, and m is the expansion coefficient correction factor. When two or more reference materials with different $\delta^{34}S$ values are analyzed, the correction for sample blank does not need to be applied if the magnitude of the blank in the samples and reference materials is identical. However, if a single reference material is analyzed, only one-point calibration is possible; therefore, a blank should be interspersed with unknowns and a blank correction should be applied.

If replicates do not agree within acceptable tolerances, they are analyzed until acceptable statistics are achieved. The computerized LIMS-LSI will not release data until the statistics are acceptable. The time required for performing the analyses is a minimum of 2 days.

No correction is made for the effect of the $\delta^{18}O$ value of the SO_2 gas on the $\delta^{34}S$ determination. Fry and others (2002) have reported that the $\delta^{34}S$ of plants, animals, and soils can be as much as 1 to 3 ‰ to positive, particularly for samples with high $n(C)/n(S)$ values. V_2O_5 is added to each sample to buffer the $\delta^{18}O$ of SO_2 gas and minimize this error (Yanagisawa and Sakai, 1983).

The systematic procedure is listed in the "Lab Procedures" binder at the RSIL. The supervisor reviews suggested changes before they are adopted.

Problematic Samples

Problematic samples include those that do not have sulfur concentration before isotopic analysis. In that case, the sulfur concentration needs to be determined by a preliminary EA analysis to be able to determine the amount of sample needed for analysis. The amounts of carbon and nitrogen may need to be established to determine whether CO_2 or N_2 gases must be diminished by helium dilution in the ConFlo before it reaches the IRMS to prevent overlap with the SO_2 gas.

Interferences

There are no known interferences with this analytical technique. However, it should be pointed out that if samples contain a mixture of different sulfur sources, reported isotope-ratio values represent the isotopic composite of the total sulfur in the sample. The $\delta^{34}S$ values of specific sulfur species cannot be provided with this analytical technique.

The $\delta^{34}S$ of plants, animals, and soils with high $n(C)/n(S)$ values may be too positive (Fry and others, 2002). To minimize this interference, V_2O_5 is added to each sample to buffer the $\delta^{18}O$ of SO_2 gas.

Troubleshooting and Bench Notes

The most common problems with this analytical technique are (1) incomplete combustion; (2) leakage in the system, particularly in the EA; and (3) insufficient linearity of the IRMS. The fact that the reaction tube temperature is only 1020 °C can create insufficient combustion. The "flash point" has a higher temperature, approximately 1800 °C (because the tin oxidation reaction is exothermic); therefore, it is critical to create a sharp, sudden, bright, single flash. This should be done by coordinating (1) the helium pressure and flow rate, (2) oxygen pressure and flow rate, and (3) sample "start time." The system is so sensitive that even a 1-second (s) difference in the sample dropping time can make a difference in the brightness of the flash, peak shape, peak area and, consequently, the $\delta^{34}S$ result. The optimum set-up parameters of the RSIL EA are as follows: cycle = 100, oxy = 60 (O_2 flow stops flowing into autosampler), sample start = 10, and sample stop = 40. The helium regulator is set for 150 kilopascals (kPa), and the helium-flow set up is 95 mL/min, which actually measures 120 mL/min at the vent. The oxygen regulator is set for 150 kPa, and the oxygen flow rate is 54 mL/min at purge. The oxygen loop is 5 mL. Oven temperature of the GC is 90 °C. Sample size commonly ranges between 0.5 and 5 mg (Révész, 1998). To improve analytical precision, 0.6 mg of V_2O_5 is added to each sample (Yanagisawa and Sakai, 1983).

Even a tiny leak can interfere with the result; it can create double peaks, an unstable analysis, and so forth. The high helium flow rate makes the system more sensitive to leaks (Bernoulli principle). Leaks can cause uncontrollable changes in the flow rate, which could prevent quantitative combustion, affecting the peak shape and peak area, and consequently, the $\delta^{34}S$ value. Ash accumulation is also critical. Although tin oxide serves as a catalyst in the reaction, it also partially "clogs" the system, interfering with the flow rate, although not necessarily in a measurable manner. However, this assumes a nonrandom interference that causes a shift in $\delta^{34}S$ value. By frequently analyzing isotopic reference materials, this shift can be corrected. However, after approximately 50 samples (one carousel), it is advisable to remove the ashes (Révész, 1998). The leak-checking procedure is given in appendix I. A leak is indicated in the IRMS when m/z 28 and 29 are high and the ratios of m/z 28, 32, and 40 (N_2, O_2, and argon) reflect atmospheric compositions. Acceptable levels for m/z 28 and 29 are approximately 0.15 and 0.015 volt, respectively, with typical IRMS settings (see "Daily Checklist" in appendix J).

Adjustment of the emission current is critical. The higher it is, the higher the signal will be, but the poorer the linearity will be. The IRMS is temperature sensitive; therefore, it must be operated in a temperature-controlled environment with all of the side panels on. The instrument is capable of measuring m/z 64 and 66 simultaneously (fig. 3).

An internal precision check needs to be conducted using the "Zero Enrichment" method (standard gas on/off 10 times), and it must give a standard deviation of 0.1 ‰ or better. The "Ratio Linearity" test must give 0.12 ‰/V or better linearity. An external precision check needs to be performed as needed, approximately every week or so; samples of the same amount (within ±0.010 mg) need to be analyzed. The 1-σ standard deviation must be 0.2 ‰ or better for n = 10. If either of these criteria has not been met, the IRMS cannot be used and must be refocused.

Maintenance and Maintenance Records

Routine maintenance is necessary for the upkeep of the IRMS vacuum system, including (1) checking the pump performance from time to time, (2) changing the pump oil, and (3) changing the oil cartridge in each turbomolecular pump at 6-month intervals. Pump conditions, including date of oil changes, problems, and repairs in the pump, are listed in the pump database (file path: LIMS C:\RSIL\vacuum pumps). Maintenance of the EA includes (1) changing the insertion tube every day (appendix K), (2) changing the water trap approximately every 200 samples (appendix L), (3) changing the reaction tube about every 200 to 250 analyses (appendix M), and (4) cleaning the sample carousel every week. The IRMS also requires a daily check (appendix J). A logbook is maintained for each IRMS, where notes of maintenance checks, history of normal settings, problems, and repairs are listed.

Maintenance of ISODAT 2.0 includes importing (transferring) results from the ISODAT 2.0 database to the LIMS-LSI computer hard disk and the Data Back-Up computer hard disk (appendix G).

Sample Retention Time and Disposal

Samples are retained in the RSIL for at least 4 months after reporting data. Samples are then discarded unless the submitter has requested that the samples be returned.

The ISODAT sample file from the IRMS computer is kept indefinitely on two different hard disks of the Data Back-Up computer. Paper reports are kept for a few weeks. Analytical results from the IRMS are transmitted to LIMS-LSI and kept indefinitely. No data are entered by hand, and no manual calculations are performed on the data.

Data Acquisition, Processing, Evaluation, Quality Control, and Quality Assurance

Laboratory Information Management System for Light Stable Isotopes (LIMS-LSI)

In the RSIL, the LIMS-LSI (Coplen, 2000) is used for data processing and evaluation. This system is a database program capable of (1) storing information about samples, (2) storing the results of mass spectrometric delta values of samples after importing them from the ISODAT database, (3) calculating analytical results using standardized algorithms stored in a database, (4) normalizing delta values using isotopic reference materials, and (5) generating templates for convenient sample placement to facilitate loading of automated mass-spectrometer sample preparation manifolds and carousels. With this system, the following are ensured: (1) quality assurance (QA), (2) laboratory efficiency, (3) reduction of workload and errors owing to the elimination of retyping of data by laboratory personnel, and (4) a decrease of errors in data reported to sample submitters. This database provides a complete record of information on how laboratory reference materials have been analyzed and provides a record of what correction factors have been used as an audit trail for the RSIL.

Quality Control (QC) Samples

Samples are prepared and analyzed in batches—a maximum of 80 daily. Each batch contains approximately six reference materials (QC samples) at the beginning: four enriched in ^{34}S, one with an intermediate value, and one depleted in ^{34}S. One set of three (depleted, intermediate, and enriched in ^{34}S) is analyzed for every 10 samples. The last analyses of a batch is that of one set of reference materials. The preparation and analytical cost of these isotopic reference materials is the responsibility of the RSIL. Both RSIL and NWQL monitor these results. Quarterly or when requested, the RSIL prepares a summary of results and provides it to NWQL.

Daily, the analyst (1) examines the computer printouts for indications of analytical problems, (2) determines the daily additive and expansion correction factors using the LIMS-LSI, and (3) applies correction factors to isotopic data. These values with dates and analysis-number ranges are recorded manually in the laboratory "δ^{34}S Correction Factor" three-ring binder even though the data are already in the LIMS-LSI. After printing the list of δ^{34}S values from the "Table of Samples in Progress," the analyst reviews the results and determines which samples need to be analyzed a third time to achieve acceptable results (that is, $\Delta\delta \leq 0.2$ ‰).

Acceptance Criteria for All QC Samples

Acceptance criteria for QC samples are the same as acceptance criteria for the other samples. The rules are as follows:

- If standard deviation is ≤0.2 ‰, use mean delta.
- If there are three or more analyses, delete the outlier and recalculate.
- If standard deviation of this recalculation is ≤0.2 ‰, use mean from this recalculation.
- If none of the above is the case, the result is not acceptable, and corrective action is required.

The RSIL estimates the expanded uncertainty ($U = 2\mu_c$) of δ^{34}S measurement results. The expanded uncertainty provides an envelope that represents a 95-percent probability of encompassing the true value that has been determined from the aggregation of measurement results over a period of time. The expanded uncertainty can be determined using the guide to the expression of uncertainty (Joint Committee for Guides in Metrology, Working Group 1 (JCGM/WG1), 2010). The application of expanded uncertainty to the reporting of stable isotope measurements is discussed by Coplen and others (2006). The estimated expanded uncertainty of δ^{34}S measurement results is ±0.4 ‰ unless otherwise specified, and this value is conservative. If any given sample were resubmitted to the RSIL for sulfur isotopic analysis, the measured value would fall within the uncertainty bounds of the previous result more than 95 percent of the time.

Corrective Action Requirements

If the analyst finds any problem with the daily reference-sample data, the analyst contacts the supervisor. This process requires an evaluation and reanalysis of certain samples to ascertain the origin of the problem.

If samples do not give satisfactory results after three or more separate analyses, the analyst averages all the data and reports the mean value. This analytical result is indicated with a comment, and the customer is advised by e-mail or by other means.

Responsible Parties for All QC/QA Functions for Procedure Covered in RSIL SOPs

The analyst, with supervisory approval, is responsible for qualifying data and notifying customers.

Data Management and Records

In addition to evaluating daily sample analyses, every week an analyst evaluates the data project by project, reports results to the customers, and files final project data reports in the laboratory "Correspondence" binder (appendix N).

Health, Safety, and Waste-Disposal Information

Applicable Health and (or) Safety Issues

Personal Protection

Safety glasses and protective gloves are recommended whenever samples are handled, especially when the samples are of biological origin. For other precautions and safety procedures, consult the Material Safety Data Sheets (MSDS), which are on file in the laboratory and at the URL *http://www.ilpi.com/msds/#Manufacturers*. This URL provides links to the MSDSs of most chemical companies.

Electrical Hazards

Electrical systems must conform to the National Electric Code, National Fire Protection Association Code (NFPA 70–1971), and the American National Standards Institute (ANSI) Code (C1–1971). For more information, consult the U.S. Geological Survey's *Safety and Environmental Health Handbook* (U.S. Geological Survey, 2002, sec. 4–4.1).

Shock hazards exist inside the instruments. Only an authorized service representative or an individual with training in electronic repair must remove panels or circuit boards where voltages are greater than 20 volts. The instruments require a third-wire protective grounding conductor. Three-to-two wire adapters are unsafe for these instruments.

Chemical Hazards

The hazardous chemicals used in the process are Cr_2O_3 and Co_3O_4, which are used in the oxidation reaction tube, and V_2O_5, which is used to boost combustion. Sometimes the reaction tube cracks, and the packing material leaks out. When this happens, protective gloves and a full-face mask are required during the cleanup process. All the materials, including gloves and any cleaning towels, need to be collected in a plastic bag and disposed of as hazardous waste. When handling V_2O_5, protective gloves and a full-face mask also are required. Upon receipt, all samples must be carefully inspected for indications of hazards.

The SO_2 reference gas is colorless and has a suffocating odor. Exposure to SO_2 can cause respiratory tract burns. Therefore, an appropriate ventilation unit must be installed in the ConFlo unit. A leak test on the tank of gas and the line carrying SO_2 to the ConFlo must be performed routinely.

Gas Cylinder Handling

Compressed gas cylinders must be handled and stored according to the U.S. Geological Survey's *Safety and Environmental Health Handbook* (U.S. Geological Survey, 2002, sec. 4–4.5.1). Each cylinder must be (1) carefully inspected when received; (2) securely fastened at all times with an approved chain assembly or belt; (3) capped at all times when not in use; (4) capped when transported; (5) transported only by a properly designed vehicle (hand truck); and (6) stored separately with other full, empty, flammable, or oxidizing tanks of gas, as appropriate.

Specific Waste-Disposal Requirements

Used reaction tubes containing Cr_2O_3, CO_3O_4, and V_2O_5 must be collected in a closed container and must be given to the safety, health, and environmental officer for disposal.

Revision History

Publication Series and Series Number: Techniques and Methods 10–C4 (Book 10, Section C, Chapter 4)

Publication Title: Determination of the δ^{34}S of Total Sulfur in Solids; RSIL Lab Code 1800

Publication Authorship: Révész, Kinga; Qi, Haiping; and Coplen, Tyler B.

Version 1.0, 2006

Version 1.1, 2007

Version 1.2, 2012

Summary of Revised Product Components

Component	Description	Last revised in publication version	Date of last revision
Title	Replaced $\delta(^{34}S/^{32}S)$ by $\delta^{34}S$ to be in accordance with recommendations of the Commission on Isotopic Abundances and Atomic Weights of the International Union of Pure and Applied Chemistry.	1.2	July 2012
Authorship	Tyler B. Coplen added as a coauthor.	1.2	July 2012
Summary of procedure	Updated text to reflect new instrumentation.	1.2	July 2012
Reporting units and operational range	Updated text to reflect expression of delta values with number ratios.	1.2	July 2012
Labware, instrumentation, and reagents	Updated text to reflect new instrumentation.	1.2	July 2012
Sample collection, preparation, analysis, retention times, and disposal	Updated text to reflect that samples are no longer submitted through the National Water Quality Laboratory (NWQL) and to document improved analytical techniques.	1.2	July 2012

Data acquisition, processing, evaluation, quality control, and quality assurance	Updated text to reflect that samples are no longer submitted through the National Water Quality Laboratory (NWQL).	1.2	July 2012
References cited	Added three references.	1.2	July 2012
Appendix A	Updated text to reflect that samples are no longer submitted through the National Water Quality Laboratory (NWQL).	1.2	July 2012
Appendix N	Updated text to reflect that samples are no longer submitted through the National Water Quality Laboratory (NWQL).	1.2	July 2012
Main text footer	Removed from document.	1.2	July 2012

References Cited

Brenna, J.T., Corso, T.N., Tobias, H.J., and Caimi, R.J., 1997, High-precision continuous-flow isotope-ratio mass spectrometry: Mass Spectrometry Reviews, v. 16, p. 227–258.

Clark, Ian, and Fritz, Peter, 1997, Environmental isotopes in hydrogeology: Boca Raton, Fla., Lewis Publishers, 328 p.

Coplen, T.B., 2000, A guide for the laboratory information management system (LIMS) for light stable isotopes—Version 7 and 8: U.S. Geological Survey Open-File Report 00–345, 121 p., accessed April 27, 2012, at http://water.usgs.gov/software/code/geochemical/lims/doc/ofr00345.pdf.

Coplen, T.B., 2011, Guidelines and recommended terms for expression of stable-isotope-ratio and gas-ratio measurement results: Rapid Communications in Mass Spectrometry, v. 25, no. 17, p. 2538–2560, accessed July 13, 2012, at http://onlinelibrary.wiley.com/doi/10.1002/rcm.5129/abstract.

Coplen, T.B., Brand, W.A., Gehre, Matthias, Gröning, Manfred, Meijer, H.A.J., Toman, Blaza, and Verkouteren, R.M., 2006, New guidelines for $\delta^{13}C$ measurements: Analytical Chemistry, v. 78, no. 7, p. 2439–2441.

Coplen, T.B., and Krouse H.R., 1998, Sulfur isotope data consistency improved: Nature, v. 392, p. 32.

Fry, Brian, Silva, S.R., Kendall, Carol, and Anderson, R.K., 2002, Oxygen isotope corrections for online $\delta^{34}S$ analysis: Rapid Communications in Mass Spectrometry, v. 16, p. 854–858.

Joint Committee for Guides in Metrology, Working Group 1 (JCGM/WG1), 2010, Evaluation of measurement data—Guide to the expression of uncertainty in measurement (known as the GUM) (2010 corrected version of the first edition of 2008): Joint Committee for Guides in Metrology [publication] JCGM 100:2008, 120 p., accessed July 16, 2012, at http://www.bipm.org/en/publications/guides/.

Krouse, H.R., and Coplen, T.B., 1997, Reporting of relative sulfur isotope-ratio data: Pure and Applied Chemistry, v. 69, p. 293–295.

Révész, Kinga, 1998, Carbon and nitrogen isotope ratios of organic and inorganic bulk samples; Instrument performance and intercalibration [abs.], *in* The 5th Canadian Continuous-Flow Isotope Ratio Mass Spectrometry Workshop, August 16–19, 1998: Ottawa, Ontario, Canada, University of Ottawa, not paged.

Robinson, B.W., 1995, Sulphur isotope standards, *in* Reference and intercomparison materials for stable isotopes of light elements: Proceedings of a consultants meeting held in Vienna, [Austria,] 1–3 December, 1993: IAEA–TECDOC–825, p. 39–45, accessed April 30, 2012, at http://www-pub.iaea.org/books/IAEABooks/5471/Reference-and-Intercomparison-Materials-for-Stable-Isotopes-of-Light-Elements.

U.S. Geological Survey, 2002, USGS handbook 445–3–H, Safety and environmental health handbook, 435 p.

Yanagisawa, F., and Sakai, H., 1983, Thermal decomposition of barium sulfate—vanadium pentoxide—silica glass mixtures for preparation of sulfur dioxide in sulfur isotope ratio measurements: Analytical Chemistry, v. 55, p. 985–987.

Appendix A. Step-by-Step Procedure to Log-In Samples to LIMS-LSI

1. For sample submitter:
   a. Download either the "Standard Submission Excel Form" or the "QWDATA Compatible Submission Excel Form" from the RSIL Web site at http://isotopes.usgs.gov/.
   b. Fill out the requested sample information.
   c. Send a diskette or CD and a hard copy along with the samples or e-mail either completed RSIL Excel worksheet to isotopes@usgs.gov.
2. For RSIL personnel:
   a. Match up information on sample bottles with the submitted "Standard Submission Excel Form" or the "QWDATA Compatible Submission Excel Form".
   b. Enter all information into LIMS-LSI by loading media containing the "Standard Submission Excel Form" or the "QWDATA Compatible Submission Excel Form". Submission date is the date samples are logged in.
   c. Use "New Project Log-in" in LIMS-LSI to assign S#s; field IDs are the Station IDs.
   d. Print out one project report and container labels (one for each sample).
   e. Put a label on a sample bottle and cross-check Field IDs between bottles and Excel worksheet data.
   f. Punch holes in the original Excel worksheet and all the project information provided and put in the "Samples-in-Progress" binder.

Appendix B. Step-by-Step Procedure to Generate an Excel Sample Workbook or to Print a Template and a Samples-to-Be-Analyzed List

Excel Sample Workbook

1. Use default worksheet-weighing template and add samples to be analyzed.
2. Fill the appropriate amount in the cells of the rows labeled "Set weight mg."
3. Print first worksheet in workbook.
4. Write the tray ID and the date on both the diskette and paper template.
5. Put them near the balance.

Template

1. Use "Print Template" in LIMS-LSI.
2. Select "appropriate template" for EA and "Delta Plus" for mass spectrometer.
3. Select "New Template" (dialog box informs you how many samples are waiting to be analyzed).
4. Click "OK."
5. Click "Print."
6. Insert diskette to receive sample headings.
7. Click "OK."
8. Write the day of the week that these samples should be analyzed on both the diskette and paper template.
9. Put them near the balance.
10. Exit LIMS-LSI.

Worksheet 1.   Weighing template.

[International reference materials are in bold font]

		1	2	3	4	5	6	7	8	9	10	11	12
A	Our Lab ID	**S-97**	**S-97**	**S-97**	**S-97**	**S-1301**	**S-1302**	S-9031	S-9031	S-9032	S-9032	S-9033	S-9033
	Set weight mg	0.3	0.3	0.3	0.3	0.3	0.3	0.3	0.3	0.3	0.3	0.3	0.3
	Actual weight, mg	0.3	0.29	0.31	0.31	0.304	0.298	0.299	0.302	0.305	0.303	0.305	0.303
	Comment												
B	Our Lab ID	S-9034	S-9034	S-9035	S-9035	**S-97**	**S-1301**	**S-1302**	S-9037	S-9037	S-9038	S-9038	S-9039
	Set weight mg	0.3	0.3	0.3	0.3	0.3	0.3	0.3	0.3	0.3	0.3	0.3	0.3
	Actual weight, mg	0.301	0.302	0.295	0.303	0.299	0.295	0.301	0.299	0.293	0.304	0.301	0.304
	Comment												
C	Our Lab ID	S-9039	S-9040	S-9040	S-9041	S-9041	**S-97**	**S-1301**	**S-1302**	S-9043	S-9043	S-9044	S-9044
	Set weight mg	0.3	0.3	0.3	0.3	0.3	0.3	0.3	0.3	0.3	0.3	0.3	0.3
	Actual weight, mg	0.301	0.301	0.295	0.306	0.297	0.302	0.3	0.305	0.304	0.297	0.304	0.301
	Comment												
D	Our Lab ID	S-9045	S-9045	S-9046	S-9046	S-9047	S-9047	**S-97**	**S-1301**	**S-1302**	S-9049	S-9049	S-9050
	Set weight mg	0.3	0.3	0.3	0.3	0.3	0.3	0.3	0.3	0.3	0.3	0.3	0.3
	Actual weight, mg	0.295	0.303	0.32	0.299	0.298	0.296	0.303	0.305	0.305	0.297	0.302	0.296
	Comment												
E	Our Lab ID	S-9050	S-9051	S-9051	S-9052	S-9052	S-9053	S-9053	**S-97**	**S-1301**	**S-1302**		
	Set weight mg	0.3	0.3	0.3	0.3	0.3	0.3	0.3	0.3	0.3	0.3		
	Actual weight, mg	296	0.296	0.295	0.299	0.305	0.305	0.299	0.302	0.297	0.304		
	Comment												

17

Worksheet 2. Samples to be analyzed.

[International reference materials are in bold font]

Line	Identifier	Port	Comment	Amount	Amt Unit
1	**S-97**	A1		0.3	mg
2	**S-97**	A2		0.29	mg
3	**S-97**	A3		0.31	mg
4	**S-97**	A4		0.31	mg
5	**S-1301**	A5		0.304	mg
6	**S-1302**	A6		0.298	mg
7	S-9031	A7		0.299	mg
8	S-9031	A8		0.302	mg
9	S-9032	A9		0.305	mg
10	S-9032	A10		0.303	mg
11	S-9033	A11		0.305	mg
12	S-9033	A12		0.303	mg
13	S-9034	B1		0.301	mg
14	S-9034	B2		0.302	mg
15	S-9035	B3		0.295	mg
16	S-9035	B4		0.303	mg
17	**S-97**	B5		0.299	mg
18	**S-1301**	B6		0.295	mg
19	**S-1302**	B7		0.301	mg
20	S-9037	B8		0.299	mg
21	S-9037	B9		0.293	mg
22	S-9038	B10		0.304	mg
23	S-9038	B11		0.301	mg
24	S-9039	B12		0.304	mg
25	S-9039	C1		0.301	mg
26	S-9040	C2		0.301	mg
27	S-9040	C3		0.295	mg
28	S-9041	C4		0.306	mg
29	S-9041	C5		0.297	mg
30	**S-97**	C6		0.302	mg
31	**S-1301**	C7		0.3	mg
32	**S-1302**	C8		0.305	mg
33	S-9043	C9		0.304	mg
34	S-9043	C10		0.297	mg
35	S-9044	C11		0.304	mg
36	S-9044	C12		0.301	mg
37	S-9045	D1		0.395	mg
38	S-9045	D2		0.303	mg
39	S-9046	D3		0.32	mg
40	S-9046	D4		0.299	mg
41	S-9047	D5		0.298	mg

Line	Identifier	Port	Comment	Amount	Amt Unit
42	S-9047	D6		0.296	mg
43	**S-97**	D7		0.303	mg
44	**S-1301**	D8		0.305	mg
45	**S-1302**	D9		0.305	mg
46	S-9049	D10		0.297	mg
47	S-9049	D11		0.302	mg
48	S-9050	D12		0.296	mg
49	S-9050	E1		0.296	mg
50	S-9051	E2		0.296	mg
51	S-9051	E3		0.295	mg
52	S-9052	E4		0.299	mg
53	S-9052	E5		0.305	mg
54	S-9053	E6		0.305	mg
55	S-9053	E7		0.299	mg
56	**S-97**	E8		0.302	mg
57	**S-1301**	E9		0.297	mg
58	**S-1302**	E10		0.304	mg

Appendix C. Step-by-Step Procedure for Weighing and Storing Samples

1. Samples should be homogenized and dried.
2. Insert template diskette in the computer connected to the microbalance.
3. Condition balance (this step should be done once a day):
    a. Place empty capsule on balance and close door. Wait for reading to stabilize (the "mg" on the display appears). Tare the balance.
    b. Remove and replace capsule and make sure the stabilized weight is 0.000 mg.
    c. Repeat these steps until the balance is stable.
4. Remove capsule from balance, add sample and weigh the filled capsule. Repeat until you have sufficient mass to yield 40 µg of sulfur.
5. Enter sample weight on the template.
6. Add 600 µg ± 100 µg of V_2O_5 to sample.
7. Fold cup; secure sample in it.
8. Repeat steps 4 to 7 for every sample.
9. Note:
    a. Do not cross-contaminate samples.
    b. Make sure the spatula and sample area are cleaned using Kimwipes between each sample.
    c. Always allow balance to stabilize before removing capsule.

Appendix D. Step-by-Step Procedure of Zero Blank Autosampler Operation

1. Before opening the lid, make sure the isolation valve (between sample chamber and reaction tube) is closed (arrow towards you) and make sure the helium purge isolation valve (on the left) is closed.
2. Open the purge vent on the top of the autosampler to vent the sample chamber.
3. Release the three fittings holding the lid closed and open the lid.
4. Place your samples and close the lid. Load 49 samples for a 50-position carousel, leaving the first hole empty.
5. Secure the lid with three bolts. Screw in all three bolts finger tight, and then lightly tighten two bolts at a time, moving around the lid until they are all completely tight to minimize stress and warping of the lid.
6. Open the helium purge isolation valve (screw down), purge the sample chamber for 5 minutes (min) at a helium flow rate of 298 mL/min. Make sure the purge vent valve (on top) is open. Press your finger on top of the vent valve for 2 s, then release; you should hear the sound of a pressure release.
7. Close the helium purge vent valve (screw down), wait for 2 min for gas pressure at sample chamber and a helium flow rate to stabilize. Close the helium purge isolation valve.
8. Open the isolation valve (arrow toward up). Wait for 2 min for stabilized baseline. With a helium carrier flow rate of 90 mL/min, one should observe *m/z* 28 on cup 1 at about 50 millivolts, *m/z* 29 on cup 2 at about 50 to 60 millivolts and *m/z* 30 on cup 3 at about 140 millivolts to 1 volt (appendix J).

Appendix E. Step-by-Step Procedure to Add Sample Information to Sequence Table

1. Insert diskette with "Weighing Template" on it.
2. Open "Sequence File" under "EA configuration."
3. Open "Weighing Template" in Excel.
4. "Copy" and "Paste" the list of Sample ID, sample weight from template to sequence table.
5. Select appropriate "method" for each sample.
6. Define reference line and blank line.
7. Start.
8. Give Folder Name: Comment
   a. Select pre: "Date."
9. Give File Name:
   a. Select pre: "Analyzes #."
   b. Select post: "Identifier 1."
   c. Select: "Print Result."
10. Click "OK."
11. Wait for first sample to be analyzed.

Appendix F. Step-by-Step Procedure to Retrieve Data from ISODAT 2.0 for LIMS-LSI and for Data Back-Up

1. For LIMS-LSI:
   a. Select "Result" in ISODAT 2.0.
   b. Select the result folder you wish to export data from.
   c. Select all the individual analyses you want to retrieve by right clicking the selected file.
   d. Select "Reprocess."
   e. Give a file name.
   f. Add export template previously designed.
   g. Open.
   h. Click "OK." (Reprocessing takes 1–2 min)
   i. Open Excel.
   j. Find the file you reprocessed (E-drive, Finnigan, ISODAT 2.0, Global, User, CONFLO II, Interface, Result: File name).
   k. Open it in Excel.
   l. Save as: "A" drive. Keep in Lotus format.
   m. Define all reference peaks by adding #1 under column "Is Ref.?."
   n. Now the file can be imported to LIMS-LSI.
2. For Data Back-Up computer:
   a. Go to Windows Explorer in ISODAT 2.0 computer.
   b. Find the drive where the data are (D).
   c. Choose: "Finnigan."
   d. Choose: "User."
   e. Choose the inlet system where you have data (Gas Bench or EA).
   f. Choose "Result folder."
   g. Transfer data to Zip disk or "memory stick."

Appendix G. Step-by-Step Procedure to Transfer Data to LIMS-LSI, to Transfer Data to Back-Up Computer, and to Reevaluate Old Data

1. Transfer data to LIMS-LSI:
   a. Start LIMS.
   b. Choose "Import Analysis."
   c. Choose mass spectrometer "Delta Plus."
   d. Click "Import."
   e. Select the file that will be imported.
   f. Select columns containing $\delta^{34}S$ data. Check the boxes to import these two columns.
   g. Click "Import." Note: Sample ID, sample weight, Peak Area, Analyses #s are automatically imported to LIMS-LSI.
2. Transfer data to Data Back-Up computer:
   a. Go to Windows Explorer in the Data Back-Up computer.
   b. Find the drive where your back ups are stored.
   c. Choose: "RSIL."
   d. Choose: "Mass. Spec. Analysis Back Up."
   e. Choose: "H." (Stands for Delta V Plus MS).
   f. Under that folder create a new folder. The name of the folder should be the date range when those analyses were done that you want to back up.
   g. Transfer data to that folder.
   h. Make back ups every 2 weeks or so, as required.
3. Reevaluate old data:
   a. Find your samples in the Data Back-Up computer.
   b. Transfer data by a Zip disk to the computer where a virtual version of ISODAT 2.0 is installed.
   c. Reevaluate your data.

Appendix H. Step-by-Step Procedure to Determine and Apply Correction Factors and Evaluate Data

1. Open "Correction Factors and Normalization Equations" in LIMS-LSI.
2. Select MS (Delta^{Plus}) and element.
3. Select "Query."
4. Double click on the last sample analyzed on that day.
5. Evaluate data of the reference materials.
6. Choose "Normalize with all References."
7. Accept "Expansion Correction and Additive Correction factors."
8. Print out correction factor sheet.
9. Report daily reference values and correction factors along with date and range of analysis number to the "EA" binder.
10. Go back to LIMS-LSI main menu by closing open windows.
11. Choose "Print Samples in Progress":
    a. Open "Sample in Progress."
    b. Choose appropriate isotope.
    c. Choose appropriate prefix (W for water, N for nitrogen, S for sulfur).
    d. Put in sample ID range from "Samples-to-Be-Analyzed" sheet. Click Print.
12. Review results; determine repeats.
13. Put repeats back to Table of Samples to Be Analyzed:
    a. Go back to LIMS-LSI main menu.
    b. Open "Print Templates."
    c. Select appropriate template name for the mass spectrometer and the samples that are to be analyzed.
    d. Find the sample in the "List of samples."
    e. Change "Repeats" from 0 to 1.
    f. Close Windows and exit LIMS-LSI.

Appendix I. Step-by-Step Procedure to Check Elemental Analyzer for Leaks

1. Close the VENT carrier output using the proper cap (provided).
2. Adjust the helium pressure to 150 kPa (regulator is on the front central panel), wait 3 min to equilibrate gas in the system.
3. Close the helium inlet valve by turning the above regulator counter clockwise.
4. If the gauge needle does not move, there is no leak.
5. If the gauge needle moves, indicating the lost of pressure, there is a leak in the system. The decreasing pressure rate accounts for the degree of leakage.
6. Locate the leak by separating and testing the system segment by segment, using the helium detector.

Appendix J. Daily Checklist

Analyst in charge: _____ Date: _____

Weekly:
1. Change the working reference gas tank and the helium gas tank if the pressure is < 500 psi. Order new ones for a spare; the reference gas is SO_2 ANHY by Matheson, the helium carrier is zero grade, and the O_2 is research grade.
2. Change water and CO_2 trap after approximately 8,000 samples, or when it is necessary.
3. Check and clean autosampler carousel.
4. Change the reaction tube after approximately 200 or 250 samples, or when it is necessary.

THESE ITEMS ARE TO BE CHECKED OFF AS YOU CHECK THEM DAILY!
1. Check He, O_2, and reference gas flow. [ ]
2. Check EA reaction-tube temperature (1020 °C). [ ]
3. Change insertion tube (ash-collector liner) in the reaction tube. [ ]
4. Check background masses. [ ]

At emission current: 1.5 milliampere

Cups	Mass	Intensity (V)	Mass	Intensity (V)	Mass	Intensity (V)	Resistor (Ω)	Capacity (pF)
1	28	0.05	44	0	64	0.002	3×10^8	680
2	29	0.05	45	0	66	0.002	3.5×10^{10}	5
3	30	0.05–0.3	46	0.002			3×10^{11}	2

5. Check peak center. [ ]
6. Analyze Ref on/off method 10 times to stabilize IRMS (standard deviation should be better than 0.1 ‰). [ ]

Optimal ConFlo Pressure Setting SO_2 (bar)	SO_2 signal (V) Mass 66 on Cup 2
0.50	3.000

27

Appendix K. Changing the Insertion Tube

The insertion tube needs to be changed before a template is run.

1. Manually close needle valve near the ion source.
2. On the menu of the Model 2500 EA under "Spc. Fun.," choose "Std By" and press "enter." The O_2 and helium gases should be turned off.
3. Unscrew the metal seals under the sample carousel to access the reaction tube.
4. Using the heat protective glove and the wire tool, remove the insertion tube from the reaction tube and place in the metal can. Be aware that the tube will be very hot!
5. Insert clean tube that has 0.5 centimeter (cm) of quartz wool packed at the bottom and tap it down with the wire tool.
6. Replace the carousel and tighten the metal seals with a wrench.
7. Turn the O_2 and the helium back on by pressing "enter" and return the screen to the temperature readout by pressing "Spc. Fun."
8. Open the needle valve.

Appendix L. Changing the Water Trap

Change the water trap after every 200 samples.

1. Retrieve quartz turnings, quartz wool, magnesium perchlorate, and a clean water trap from Delta V Plus supplies.
2. Under a hood, pack one end of the clean water trap with 1 cm of quartz wool.
3. On a piece of wax paper, create a mix by volume of approximately 70-percent magnesium perchlorate and 30-percent quartz turnings. Pack the mix into the water trap tube, leaving 1 cm of empty space.
4. Pack the open end of the water trap with quartz wool.
5. Unscrew the red plastic ends from the used water trap and replace it with the clean water trap; be sure to replace the rubber O-rings on the ends of the clean water trap and tighten the red ends.
6. Insert the plug into the vent and watch for any decrease in pressure to check for a leak in the system. If everything looks acceptable after 5 min, replace the vent line.
7. Under the "Spc. Fun." menu of the model 2500 EA, choose "CNT" (counter) and zero "D" to set back the water-trap counter.
8. Clean the water trap by pushing all the mixture out of the tube into the trash can and use a clean cotton swab to wipe out the inside of the tube. Rinse the tube with DIW, and dry it in the oven. Return this tube to the water trap bag in the drawer.

Appendix M. Changing the Reaction Tube

The reaction tube should be changed approximately every 200 to 250 samples.

1. Manually close the needle valve near the ion source.
2. Turn the Model 2500 EA to "Std By" with both gases off.
3. Lower the reaction tube temperature to less than 900 °C.
4. Retrieve the reaction tube from the cabinet.
5. Insert the insertion tube into the top of the reaction tube; there will already be a plastic tube in the reaction tube that should be removed. The bottom of the reaction tube has copper in it.
6. Place a black O-ring on the top of the reaction tube. Make sure that all O-ring connections are clean and free of any dust.
7. Loosen the top metal seal and remove the bottom metal seal. Go back and completely remove the top seal. It will be very hot! Use oven gloves!
8. Remove quartz reaction tube and place in metal can, place clean reaction tube in column. Make sure the O-ring is clean; then replace the top.
9. On the bottom, put on the metal piece with the metal O-ring first and then add the clean rubber O-ring. Raise the bottom portion and screw together. Then tighten the connection with wrenches and put the metal support back in place.
10. Turn the gases back on; then, go to "CNT" in the menu under "Spc. Fun." and zero "B," the oxidation reaction tube as well as "A" and "C," to set back the counter.
11. Insert the plug into the vent and watch for any decrease in the pressure to check for a leak in the system. If everything looks okay after 3 min, replace the vent line.
12. Open the needle valve.

Appendix N. Step-by-Step Procedure to Report Data

1. Open "Store Samples in Progress" in LIMS-LSI.
2. Choose the appropriate isotope.
3. Choose sample ID range from "Sample in Progress" print out.
4. Store data.
5. Go back to the main menu of LIMS-LSI.
6. Open "Project" and find the appropriate project in the list.
7. Select "Print Report" and check whether the project report contains all the results. If not, search for the missing results in the database.
8. Select "Results," transfer data in Excel format or (and) text format to a diskette, and report data to customer through e-mail.
9. Click "Print Report" to print a project report and put it in the "Correspondence" binder along with all the other documents in the "Samples In Progress" binder that are related to this project.